遍地是金
Gold Everywhere

Gunter Pauli

冈特·鲍利 著

唐继荣 译

丛书编委会

主　任：贾　峰

副主任：何家振　郑立明

委　员：牛玲娟　李原原　李曙东　吴建民　彭　勇
　　　　冯　缨　靳增江

丛书出版委员会

主　任：段学俭

副主任：匡志强　张　蓉

成　员：叶　刚　李晓梅　魏　来　徐雅清　田振军
　　　　蔡雩奇

特别感谢以下热心人士对译稿润色工作的支持：

姜竹青　韩　笑　杨　爽　周依奇　于　哲　阳平坚
李雪红　汪　楠　单　威　查振旺　李海红　姚爱静
朱　国　彭　江　于洪英　隋淑光　严　岷

目录

遍地是金	4
你知道吗?	22
想一想	26
自己动手!	27
学科知识	28
情感智慧	29
艺术	29
思维拓展	30
动手能力	30
故事灵感来自	31

Contents

Gold Everywhere	4
Did you know?	22
Think about it	26
Do it yourself!	27
Academic Knowledge	28
Emotional Intelligence	29
The Arts	29
Systems: Making the Connections	30
Capacity to Implement	30
This fable is inspired by	31

两只海鸥正在垃圾堆里寻找腐肉作为食物，并为这些破烂来自哪里感到好奇。

"我们过去在这里能找到好食物，"老海鸥哀叹道，"但现在人类用有机废物制造肥料。一些人甚至用剩下的咖啡渣种蘑菇！"

Two seagulls are scavenging a rubbish dump and wondering where all the junk comes from.

"We used to find good food here," laments the older seagull, "but now humans are making fertilizer from organic waste. Some are even farming mushrooms on their leftover coffee grounds!"

两只海鸥在寻找腐肉

Two seagulls are scavenging

扔掉各种各样的东西

Throwing away all kinds of things

"对我们来说,那可不是个好消息。这些天他们扔掉各种各样的东西——在我的一生中,还从未看到过这么多的破手机。"另一只海鸥抱怨道。

"我甚至不得不把我的食物从旧电池上咬下来!"

"That's bad news for us. These days they're throwing away all kinds of things – I've never seen so many broken cellphones in my life." observes the other seagull.

"I've even been forced to pick my food off old batteries!"

"这些人类宣称要回收电池。那些又大又重的黑色电池已经没有了，但我过去从没见过这么多小电池。它们的味道真糟糕！"

"我听说人类打算烧掉所有废物，只把剩下的灰烬扔到垃圾场。"老海鸥难过地摇摇头。

"那就意味着我们将没有食物啦！"年轻的海鸥惊叫。"难道人类没有意识到，燃烧垃圾虽然节约空间，但会把所有的东西变得有毒？"

"These humans claim to recycle batteries. Those big, heavy black ones are gone, but I've never seen so many tiny ones. They taste awful!"

"I hear they're planning to burn all the waste, and only throw the leftover ash onto this dump." The old seagull shakes his head sadly.

"That means there'll be no more food for us!" the younger seagull exclaims. "Don't they realise that burning rubbish saves space, but turns everything toxic?"

烧掉所有废物

Burn all the waste

把金属，甚至是黄金，转变为气体

Turn solid metal - even gold - into gas

"他们不可能这么愚蠢,不会只是担心空间够不够!但愿在用火烧掉这里之前,所有计算机、电话机、电视、打印机和电冰箱都先被拿走了。"

"我早就听说,他们将在一个巨大的炉子里烧掉所有的电子垃圾。"年轻的海鸥说。

"我也听说了!这个做法是先用强酸处理,然后把金属气化。这些人类懂得怎样把金属,甚至是黄金,转变为气体。"

"They can't be so dumb as to only be worrying about space! Let's hope that all the computers, phones, TVs, printers and refrigerators are taken out of here before they try burn it."

"There's a lot of talk that they're going to burn all electronic waste in a huge furnace," says the younger seagull.

"I've heard! The idea is to use strong acids and then to evaporate metals. These humans know how to turn solid metal – even gold – into gas."

"想一想那该多热呀!我曾经试着在妈妈的煎锅里煎一些电池……那真是个坏主意。"年轻的海鸥看上去有些负罪感。

"那太危险了!"老海鸥训道,"你不该那样干,它会让大家永远生病的!"

"Imagine how hot that must be! I once tried to fry some batteries in my mom's frying pan … That was a bad idea." The young seagull looks guilty.

"That is so dangerous!" scolds the older seagull. "You should never do that – it could make everyone sick forever!"

会让大家永远生病

Could make everyone sick forever

花几十年来收拾这个烂摊子

Decades to clean up this mess

"我肯定不会。但为什么他们一开始就要制造所有这些有毒的填充物呢?"

海鸥叹了口气,说道:"即使他们停止制造所有的电子垃圾,即使他们终于开始重新利用它们,仍然要花几十年来收拾这个烂摊子。"

"你知道吗?一吨旧手机含有的黄金和铂金含量比一吨开采出来的矿石中的含量还多。"

"Sure, but why do they make all this toxic stuff in the first place?"

The seagull sighs. "Even if they stop making all their electronic junk, it's still going to take decades to clean up this mess – even if they finally start reusing it."

"Did you know there is more gold and platinum in a ton of old cellphones than in a ton of ore excavated from a mine?"

"嗯，你知道吗？在现在遍布全世界城乡各地的矿渣堆里还有大量贵金属，造成了许多健康问题。"

"嗯，你知道吗？旧的深矿井还灌满了饮用水，但这些水现在只用来稀释被污染的水。"

"你是说饮用水中含有微小的金片？这是惊人的浪费。"

"Well, did you know that there's still lots of precious metals in the mine dumps that now litter the world's cities and towns, causing health problems?"

"Well, did you know that the old, deep mines are full of drinking water that is now only being used to dilute polluted water?"

"You mean drinking water with minute flakes of gold in it? Amazing that it's wasted."

深矿井灌满了饮用水

Deep mines are full of drinking water

在矿井周围被污染的土地上种庄稼、用来制造生物燃料……

Plant crops on the polluted land around the mines to make biofuels...

"是呀！你知道吗？在老矿井有水从几千米高空落下，所产生的热可以用来发电。"

"嗯，你知道吗，人类可以在矿井周围被污染的土地上种庄稼，用来制造生物燃料。那才是真正的金子！"

"嗯，你还知道吗？这些废料堆将成为未来的金矿。没有人会再去挖洞了！与此相反，人类将对这些剩下的山体进行开发。"

"Yes, and did you know that water falling thousands of metres in the heat of an old mine can be used to generate power?"

"Well, did you know that humans could plant crops on the polluted land around the mines to make biofuels? Now that's real gold!"

"Well, did you know that these waste dumps will be the mines of the future? No one will dig holes any more – instead, humans will excavate these leftover hills."

"好吧！我的确懂得，我们身边遍地是金。"

"人类认为采矿业没有未来，许多人将失去工作，这是不是让人感到惊叹？我希望下辈子回来时不再是海鸥，而是一位现代矿工！"

……这仅仅是开始！……

"Well, what I do know is that there's gold all around us."

"Isn't it amazing that humans believe that there's no future in mining and that many will lose their jobs. I hope that in a next life I won't come back as a seagull, but as a modern-day miner!"

... AND IT HAS ONLY JUST BEGUN!...

……这仅仅是开始！……

...AND IT HAS ONLY JUST BEGUN!...

Did You Know?
你知道吗？

The largest landfill of the United States is as tall as the skyscrapers of Los Angeles, with 1500 trucks delivering 12 000 tons of waste per day.

美国最大的垃圾填埋场有洛杉矶的摩天大楼那么高，每天有1500辆卡车运来1.2万吨的废物。

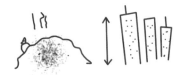

The USA alone accumulates 250 million tons of garbage a year. The main problem is not the piles of waste but the leachate – a noxious brew of chemicals that leaks into the groundwater.

仅在美国，每年积攒的厨余垃圾（生活垃圾）就有2.5亿吨。主要问题不在于废物的堆积，而是渗滤液，也就是由化学品混合后产生并泄漏到地表水中的有毒物质。

In Germany and Scandinavia, landfill is the last resort: today, only 5% of waste ends up in landfills as the majority is incinerated and recycled.

在德国和一些北欧国家，垃圾填埋场是最后的手段了。如今只有5%的废物最终到达垃圾填埋场，而绝大部分被焚化或回收了。

Americans discard 30 million computers each year, and Europeans dispose of 100 million phones each year. E-waste will increase by a factor of 5 in the next decade.

美国人每年丢弃3千万台计算机，而欧洲每年会处理掉1亿部电话机。在未来十年，电子垃圾将增加5倍。

A ton of ore from a goldmine yields 5g of gold; a ton of discarded mobile phones could yield 150g, along with 100kg of copper and 3kg of silver.

一吨从金矿开采出来的矿石能产出5克黄金；一吨被丢弃的手机能产出150克黄金，以及100千克铜和3千克银。

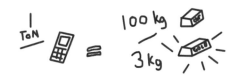

Tailings (the residue of ore) still contains gold, and can be used to make bricks. The leftovers are good for making of stone paper, the production of which needs no cellulose and uses no water.

尾矿（矿石提炼后的残留物）仍旧含有黄金，也能用于制造砖头。矿石边角料也可以用来制造石头纸，后者的生产不需要纤维素，也不需要消耗水。

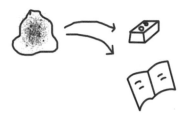

Deep mine shafts fill with warm water that can generate power: for example, in mines in South Africa, water flows downwards for 4000 metres.

深矿井充满能用来发电的温水。例如在南非的矿山中，水的落差达4000米。

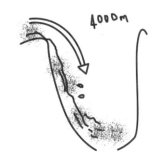

A particle accelerator boosts beams of carbon at the speed of light, using vast supplies of energy to convert lead into gold.

利用巨大的能量，粒子加速器能把碳离子束加速到光速，将铅元素转化为金元素。

Think About It
想一想

Would you rather mine gold from the Earth's crust or from your cellphone?

从地壳开采金矿或者从你的手机回收黄金，你选择哪一样？

可以从周围的废物堆和旧矿山赚钱，你同意吗？

Do you agree that there is money to be made around waste dumps and old mines?

If you were able to come back to the Earth in a distant future, what job would you like to have?

如果你能在遥远的未来回到地球，你希望从事什么工作？

把金属加热和气化像是一个危险的工作岗位吗？

Does the heating and gasification of metals seem like a dangerous job?

Do It Yourself
自己动手

Make an inventory of all the electronic equipment around your house. Everything needs to go on the list, including a small lantern that has a computer chip in it, your electric tooth brush and the quartz watch that has a button battery. Then ask how many of your appliances rely on batteries? Most importantly, what are you doing with the old batteries? What happens to the electronics at the end of their life? Who is in charge of recycling them? Check the weight of everything you have in the house and then estimate how much copper, silver and gold may be hidden in them. Urban mining may very well start at home.

列一份你房子周围的电子设备目录清单，涵盖所有的相关设备，包括有内置计算机芯片的小电灯、电动牙刷和有纽扣电池的石英手表。然后提问，这些家用电器中有多少依赖电池？更重要的是，你是怎样处理旧电池的？在那些电子产品的使用寿命终结时，你是怎么处理它们的？谁来再循环利用它们？检查一下房子里所有电器的重量，估计它们中隐藏有多少铜、银和金。都市采矿业完全可以从家开始！

TEACHER AND PARENT GUIDE

学科知识
Academic Knowledge

生物学	耐金属贪铜菌以纳米粒的形式在体内积聚金元素；某些细菌通过排出的代尔夫肌动蛋白（与金离子）发生化学反应，把溶液中的金元素沉淀出来。
化 学	电子废物不仅包含重金属，还包括有毒的溴系阻燃剂和塑料添加剂；樱桃和苹果种子含有（少量的）天然氰化物。
物 理	金属在高度真空的条件下蒸发；胶质银和胶质金（纳米金）是存在于液体中的微观粒子；金元素具有弹性光散射特性，表现为红色、蓝色或金色；粒子加速器在获得巨大能量供应下，能把铅元素转变为金元素；金元素不生锈或变色。
工程学	干电池或纽扣电池能生产出有价值的可回收金属；黄金可以先被蒸发，然后沉积为一层薄膜；尾矿坝是采矿废物被永久保存的场所；在公元前1200年发明出来、用于生产珠宝和艺术品的"失蜡"技术（即熔模技术），至今还被采用。
经济学	采用金本位制货币体系的国家，它们的货币与黄金重量挂钩；淘金热是一种早期投机形式：只有三分之一的投机者挖到黄金，却需要5年时间才能抵消到到达淘金地的花费；供应的重要性：黄金在16世纪时比白银贵15倍，如今比白银贵50倍。
伦理学	当在我们的电子垃圾中就含有黄金时，我们怎么能继续开采金矿，把地球弄得满是伤疤，造成环境污染并制造健康危机呢？
历 史	埃及人在公元前3600年就融化了黄金；来自土耳其的吕底亚人在公元前564年首次铸造了金币，而中国在同一年有了首枚方形金币；黄金艺术品是波斯的祆教信仰的一部分。
地 理	世界上最大的金矿在印度尼西亚、南非和巴布亚新几内亚；世界上最大的垃圾填埋场很可能是太平洋，而陆地上的最大垃圾填埋场位于韩国。
数 学	在黄金价格和政府债务之间的相关性方面，趋势有联系是显而易见的。
生活方式	电子产品正在被快速消耗，却没有人关心它产生的废物，也没有从这些废物中回收那些贵金属。
社会学	在很多文化中，黄金已经被用作交换媒介、计价单位和价值符号；黄金象征着财富，而且在社会上是权力的保障。
心理学	黄金与男性能量和太阳的力量相联系，白银与女性能量和月亮的能量相联系；金色是获胜的颜色，被看作是乐观和积极的颜色。
系统论	我们处置这么多废物，但在资源回收上所做甚少。我们太专注于我们的日常问题，错过了能带来收入和工作的机遇。

教师与家长指南

情感智慧
Emotional Intelligence

海鸥

海鸥很聪明，消息灵通。他们对没有过去那样多食物的现状感到焦虑，而他们的不适也与由电池释放的重金属有关。海鸥已经就谁将影响他们的未来进行了一次公开讨论：如果所有垃圾都被焚烧，他们就没有吃的了！海鸥展示了逻辑推理能力，不相信人类不愿意分享他们的见解和观点。海鸥精通对垃圾管理的技术理解，不带骄傲或争斗性地分享他们的知识。他们在沟通时毫无拘束，表达他们的意见时既无保留又不过分。海鸥对时间和空间很敏感，同时维持一种全球范围的视角。海鸥内在天性积极乐观，即便是处于压倒性的不利境地，也通过揭示没有被利用的机遇进行竞赛，鼓励彼此采取建设性的态度。这种积极的氛围激励其中一只海鸥想成为新时代的矿工。

艺术
The Arts

你能找到一些旧的印制电路板（PCB）吗？把这种带有许多晶体管的绿色面板设想成拼图的一部分。你既可以把它看作废物，也可以把它视为艺术品。利用你的想象力，把印制电路板创造成为一件艺术品。一定要把图片发给我们哦！我们很可能在下一版的《冈特生态童书》发表它。

TEACHER AND PARENT GUIDE

思维拓展
Systems: Making the Connections

我们不是消费者社会，而是用后即弃型社会。在进入我们经济体系的所有原材料中，平均只有10%是被实际消费掉的。我们真正消费的是既定企业成长战略的一部分，称为"计划性报废"，在这种战略中，功能性产品注定会由于软件更新、性能的边际改进或设计的改变而发生功能性失调。这些产品大多数属于电子产品，包括配有铜线和显示单元的印制电路板。产品的电子化已经降低了零件的数量，提高了装配速度，降低了生产和维护的成本；这也意味着电子元件现在已经渗透到我们的家庭、办公室和工厂。当这些数字产品被抛弃时，同样意味着主要为聚合物形式的电子废物的增长。大部分电子废物没有得到循环利用。电子废品的循环利用自身是高度集中的，需要大规模运送，而这类金属和塑料的混合物的分离过程需要很多能量和化学品。新的机遇也在显现，能完善循环利用过程，消除酸浸污染。即便这些良性技术（例如螯合作用）已经是众所周知并得到了坚实的证明，却很少有工业企业去推行，而是情愿坚持那些"别人都在干"的方式。现有企业关注已经感知到与采用其他可选技术相关的风险，却没有评估所有（明显）的机遇。在采矿业方面（来自垃圾填埋场和露天矿或绿地竖井矿的都市型开采），需要突破这种普遍的无知。对于像开采业这样的工业来说，有这么多的积极机遇可供利用，能很容易成为对社会有积极贡献的企业。

动手能力
Capacity to Implement

利用谷歌地球（Google Earth）作为工具，近距离观察一些旧矿山，注意是竖井矿而不是露天矿，特别是那些运行至少达50年的老矿。列出至少10个由矿山造成的问题，包括矿产公司需要应付的负面议题。然后列出一份你在某处旧金矿现场观察到的机遇清单。系统地阐述你的观点：采矿业有未来吗？你认为采矿业可以激发年轻工人和工程师去寻求光明的职业生涯吗？

教师与家长指南

故事灵感来自

马克·库蒂法尼
Mark Cutifani

马克·库蒂法尼出生于澳大利亚悉尼市外的伍伦贡。他大学毕业时获得采矿工程专业的学位，并在澳大利亚的采矿业开始他的职业生涯。经过几年在加拿大的杰出工作，他先是成为世界上第二大黄金开采公司的首席执行官，后来在一家位居世界上最大的多元化矿业公司之一的公司担任首席执行官，后者几乎开采从铂、钻石到铜、铁和煤的所有矿产。他早就意识到采矿业面临的挑战，通过探索新的产业选择把管理转向更大的价值产出，致力于给这一产业提供更可持续的发展基础。

更多资讯

http://www.scientificamerican.com/article/fact-or-fiction-lead-can-be-turned-into-gold/

图书在版编目（CIP）数据

遍地是金：汉英对照 /（比）鲍利著；唐继荣译 . -- 上海：学林出版社，2015.6
（冈特生态童书 . 第2辑）
ISBN 978-7-5486-0876-9

Ⅰ.①遍… Ⅱ.①鲍… ②唐… Ⅲ.①生态环境-环境保护-儿童读物-汉、英 Ⅳ.① X171.1-49

中国版本图书馆 CIP 数据核字 (2015) 第 092487 号

© 2015 Gunter Pauli
著作权合同登记号　图字 09-2015-446 号

冈特生态童书

遍地是金

作　　者——	冈特·鲍利	
译　　者——	唐继荣	
策　　划——	匡志强	
责任编辑——	匡志强　蔡雩奇	
装帧设计——	魏　来	
出　　版——	上海世纪出版股份有限公司 学林出版社	
	地　址：上海钦州南路81号　电话／传真：021-64515005	
	网址：www.xuelinpress.com	
发　　行——	上海世纪出版股份有限公司发行中心	
	（上海福建中路193号　网址：www.ewen.co）	
印　　刷——	上海图宇印刷有限公司	
开　　本——	710×1020　1/16	
印　　张——	2	
字　　数——	5万	
版　　次——	2015年6月第1版	
	2015年6月第1次印刷	
书　　号——	ISBN 978-7-5486-0876-9/G·325	
定　　价——	10.00元	

（如发生印刷、装订质量问题，读者可向工厂调换）